Live Oak/Milstar Complex and Carpet Service Center LaGrange, Georgia

Investigated by: Thomas H. Miller, P.E.

This is Report 086 of the Major Fires Investigation Project conducted by Varley-Campbell and Associates, Inc./TriData Corporation under contract EMW-94-C-4423 to the United States Fire Administration, Federal Emergency Management Agency.

 FEMA

Department of Homeland Security
United States Fire Administration
National Fire Data Center

U.S. Fire Administration Fire Investigations Program

The U.S. Fire Administration develops reports on selected major fires throughout the country. The fires usually involve multiple deaths or a large loss of property. But the primary criterion for deciding to do a report is whether it will result in significant "lessons learned." In some cases these lessons bring to light new knowledge about fire--the effect of building construction or contents, human behavior in fire, etc. In other cases, the lessons are not new but are serious enough to highlight once again, with yet another fire tragedy report. In some cases, special reports are developed to discuss events, drills, or new technologies which are of interest to the fire service.

The reports are sent to fire magazines and are distributed at National and Regional fire meetings. The International Association of Fire Chiefs assists the USFA in disseminating the findings throughout the fire service. On a continuing basis the reports are available on request from the USFA; announcements of their availability are published widely in fire journals and newsletters.

This body of work provides detailed information on the nature of the fire problem for policymakers who must decide on allocations of resources between fire and other pressing problems, and within the fire service to improve codes and code enforcement, training, public fire education, building technology, and other related areas.

The Fire Administration, which has no regulatory authority, sends an experienced fire investigator into a community after a major incident only after having conferred with the local fire authorities to insure that the assistance and presence of the USFA would be supportive and would in no way interfere with any review of the incident they are themselves conducting. The intent is not to arrive during the event or even immediately after, but rather after the dust settles, so that a complete and objective review of all the important aspects of the incident can be made. Local authorities review the USFA's report while it is in draft. The USFA investigator or team is available to local authorities should they wish to request technical assistance for their own investigation.

This report and its recommendations were developed by USFA staff and by Varley-Campbell & Associates, Inc. Miami and Chicago, its staff and consultants, who are under contract to assist the Fire Administration in carrying out the Fire Reports Program.

The U.S. Fire Administration greatly appreciates the cooperation received from Fire Chief Chris Smith and many of the officers and firefighters of the LaGrange Fire Department. Appreciation also goes to Deputy Chief of Suppression John L. Morris and Captain Randall H. Cash, Prevention.

For additional copies of this report write to the U.S. Fire Administration, 16825 South Seton Avenue, Emmitsburg, Maryland 21727. The report is available on the Administration's Web site at http://www.usfa.dhs.gov/

U.S. Fire Administration

Mission Statement

As an entity of the Department of Homeland Security, the mission of the USFA is to reduce life and economic losses due to fire and related emergencies, through leadership, advocacy, coordination, and support. We serve the Nation independently, in coordination with other Federal agencies, and in partnership with fire protection and emergency service communities. With a commitment to excellence, we provide public education, training, technology, and data initiatives.

 FEMA

TABLE OF CONTENTS

Live Oak/Milstar Complex and Carpet Service Center
LaGrange, Georgia
January 31, 1995

Milliken & Company
300 Lukken Industrial Drive West
LaGrange, Georgia

Local Contact: Chris Smith, Chief
John L. Morris, Deputy Chief, Suppression
Randall H. Cash, Captain, Prevention
LaGrange Fire Department
301 Main Street
LaGrange, Georgia 30240

Dean Jackson, Engineering Services Manager
Milliken & Company
Lukken Industrial Drive West
LaGrange, Georgia 30240

OVERVIEW

An afternoon fire on Tuesday, January 31, 1995 destroyed Milliken & Company's Live Oak/Milstar Complex and Carpet Service Center. This was an approximately 600,000 square feet, fully sprinklered carpet manufacturing, warehousing, cutting and distribution facility. The fire began shortly before 2:00 p.m. in the northeast corner of the primarily single story structure in the carpet manufacturing area. The fire initially involved a laminating machine, which attaches carpet to different types of backing. The fire quickly overtaxed the wet pipe automatic sprinkler protection resulting in collapse of the roof and major structural elements within 10 to 20 minutes of the fire's start. The direct dollar loss to building and contents has been estimated at over $190 million and the total loss at over $400 million.

The complex started as a carpet warehouse and distribution center in 1968 and was expanded in several major stages through 1973. (See Appendix A for building diagram). After 1973, interior revisions to production and processing machinery continued with a major production renovation in the area of fire origin reportedly completed in 1991. The cause of this fire is attributed to a rotating coupling failure that released combustible hot oil from a closed loop heat transfer system. (See Appendix B for information on heat transfer systems). This combustible liquid ignited, resulting in an intense, three-dimensional fire. A three-dimensional fire includes the liquid burning in the two-dimensional spill pool on the floor plus the burning liquid falling from the coupling above the pool.

Two 8-inch automatic sprinkler risers were located near the point of origin. These risers supplied automatic sprinklers located below the roof in the area of fire origin and an area immediately adjacent. These automatic sprinkler risers or the large, horizontal supply mains connected to them were likely destroyed by the roof collapse early in the fire, interrupting the water supply to the sprinklers and resulting in additional structural collapse. The failed risers also drained water from other sprinkler systems around the area of origin. Fire protection water supplies are usually not designed for multiple sprinkler systems operating at the same time. Water flowed from broken automatic sprinkler mains and risers for over seven hours during the fire suppression efforts, severely taxing the municipal water system in the area surrounding the plant.

As new buildings were added to the complex, the former masonry exterior walls were used as interior fire separations. Rolling type fire doors were installed in the large openings created in these walls to connect the buildings. Several large conveyor openings were identified as being protected by deluge sprinklers or spray nozzles. Water supplies for this deluge protection came from the same water supply system that supplied the automatic sprinklers and which was destroyed early in the fire. Reports from interior sector officers and firefighters indicated that nearly all of the fire doors failed to close automatically and could not be closed manually. Fire door maintenance deficiencies reportedly had been brought to the attention of plant management before the fire.

The initial fire in this highly protected risk (HPR) property involved a fuel that is difficult to extinguish with water. The failure to control the fire resulted in an early structural collapse which destroyed large automatic sprinkler mains and risers, which overwhelmed the plant's fire protection water supply and the municipal water system. Delayed alarm, sparse pre-incident planning, and alleged deficient fire door maintenance also contributed to the total loss of the complex. Due to restricted site access and the limited ability to confirm the details given by others, this report is based upon the best information available.

KEY ISSUES

Issues	Comments
Building Fire Protection	The automatic sprinkler system was overwhelmed by the fire's size and fuel supply. Early failure of two large risers/supply mains diverted water from sprinklers over the fire. Failed automatic sprinkler systems were not promptly shut down to direct the water to viable systems.
Water Supply	The large on-site stored water supply was lost early in the fire due to failure of the diesel engine driven fire pumps. The city water system flowed directly into the plant's system and was taxed by flow from over ten 8-inch collapsed automatic sprinkler risers early in the fire.
Process Controls	Heat transfer system controls and emergency procedures may not have been sufficient to prevent the combustible heat transfer fluid from continuing to feed the fire.
Fire Door Maintenance	Many rolling steel fire doors installed in masonry walls between additions to the Complex reportedly failed to close automatically or manually. The principle failure mechanism was identified by fire department and insurance company representatives as pinched together guide tracks at the edges of these doors. These sources also indicated that fire door and other maintenance deficiencies had been discussed with the plant management prior to the fire.

Issues	Comments
Delayed Alarm	The direct fire alarm connection to the county 9-1-1 emergency center was placed out-of-service by the plant. The purpose of doing this was to perform maintenance on a sprinkler system riser that was in a section of the plant away from the point of fire origin. The first telephone report of the fire to the 9-1-1 emergency center was from an in-plant location near the loading dock. This was followed several minutes later by a telephoned report from the plant security office. This demonstrated a lack of coordination between plant security and maintenance functions which should have determined that the initial water flow alarm was not from the sprinkler riser being repaired.
Conveyor Openings	Several large conveyor openings through fire separation walls were protected by deluge sprinklers. The loss of water supply to these sprinklers early in the fire rendered this protection inoperative. Smoke, hot gases and eventually the fire were able to travel through these openings because of the lack of water spray and fire doors.
Pre-Incident Planning	The fire department reported visiting the plant in early 1994 as a new carpet manufacturing process was starting-up and that the Fire Prevention Bureau was well equipped with drawings, blueprints and pre-fire plans. Company familiarization tours and comprehensive fire prevention visits probably had not been conducted since that time. Plant knowledge and familiarization is a continuous undertaking for all fire department members. At a senior level, planning should also include an analysis of what might occur if a fire protection feature, such as the automatic sprinklers, fails.
Multiple Alarms	The Milliken fire was a third alarm incident that included recall of off-duty LaGrange firefighters and mutual aid from other towns and county resources. The LaGrange Fire Department has procedural guidelines for equipment responses to confirmed structure fires through the third alarm which would provide a total of 12 engines, 4 trucks and 3 squads. However, the guidelines do not identify the source and units of mutual aid equipment responding on the various alarms. Based upon interviews, the number of fire apparatus that were operating at the scene was less than the preplanned response.

BUILDING HISTORY AND OCCUPANCY

The Live Oak/Milstar Complex and Carpet Service Center is located on the south side of LaGrange on Lukken Industrial Drive West (See Appendix A). The area is a combination of established and developing industrial sites along a primary east-west road. The area is either industrial building or open land except for a group of single story wood frame dwellings immediately to the north. According to LaGrange Fire Department's Public Information Officer, approximately 50 homes were evacuated during the fire and for about three days thereafter.

The first building on the site was constructed in 1968 as a 270 feet by 420 feet single-story with partial mezzanine, Carpet Service Center. This facility received carpet manufactured elsewhere, stored it in carpet roll racks, and cut carpet rolls to size for orders. This function continued in operation at the time of the fire although on a reportedly smaller scale.

In 1970, the Service Center was expanded to the north and to the east by a roughly 118,000 square feet addition. At the time of the fire, this area contained carpet roll storage in racks and cardboard boxed carpet tile stored in racks and in the aisles. The former exterior walls became fire separations between the original building and the 1970 addition. Large lift truck door openings were indicated to be protected by single or double rolling fire doors.

The Live Oak/Milstar Complex is the manufacturing plant and was constructed in 1972 by an addition of about 280,000 square feet to the then existing complex's north and east sides. Besides manufacturing, the 1972 addition provided more storage area, offices and support facilities, such as testing laboratory and maintenance shops.

In 1973, a 24,000 square feet addition filled an open area on the south side of the complex. Originally the addition was for use as warehouse and shipping space. At the time of the fire, this area was being used for the cutting and packaging of carpet tile squares which was the plant's product. A small shipping/staging area of about 8,000 square feet was added on the complex's west side near the northwest corner in 1986. The addition contained numerous truck dock doors and was separated from the warehouse area by the former masonry outside wall with openings protected by rolling fire doors.

A thermal transfer fluid system was installed in the northeast corner of the manufacturing plant. (See Appendix B for background information on these systems.) Thermal transfer fluid systems are often referred to as "hot oil" systems. The system at the Live Oak complex consisted of a main fluid supply and return loop with three separate heat consumer or load loops. The main loop consisted of two direct fired heaters with two circulating fluid pumps located outside the northeast corner of the complex. This equipment was located under a protective metal canopy. Milliken indicated that the main fluid loop operated at 50 to 60 pounds per square inch (psi) pressure with an estimated flow rate of 200 gallons per minute (gpm). A 500 gallon capacity expansion tank was located above the roof of the two-story mixing and blending building and reportedly was connected to the main fluid circulating loop. It was estimated that the total system – main loop and the three load loops – contained about 3,000 gallons of hot oil (heat transfer fluid).

Each of the heat consumer or load loops were described as having independent circulating pumps which moved the hot oil around the loops. Each load loop had a temperature control value which admitted heated oil from the main loop and returned cooled oil to the main loop's return path for reheating. The circulating flow rates of the load loops were not identified but operating pressures were thought to be similar to the main loop. Control and shut-down of each load loop was reported to be independent of the main loop.

According to the LaGrange Fire Department, the manufacturing equipment in the area of fire origin was last modified in 1994. This equipment was identified as "Laminating Range Number 6" and it attached different types of backing materials to rolls of carpeting. The range is an assembly of a continuous adhesive coater and heated rollers through which the carpet product passes. The laminating range consisted of two equipment levels and was 15 to 20 feet wide and 100 to 150-feet long. The heated proprietary adhesive was described by Milliken as a molten plasticizer rather than a solvent based material. It was blended in heated mixing vats located in an area adjacent to the range in the northeast corner of the complex. At the time of the fire, this laminating range was processing carpeting and a cushion type backing material described as a roll of felt with a urethane foam cushion.

The roll of backing material and roll of carpet are each coated on one side with the heated plasticizer adhesive in a continuous process. The two components are joined as they pass through rollers heated by the hot oil. Operating temperature of the rollers was about 380 degrees Fahrenheit. Each heated roller was supplied with hot oil through a 1-1/2-inch diameter rotary coupling. Another rotating coupling returns the oil to the closed loop circulating system. Each coupling was supplied from a 3-inch diameter insulated pipe with a similar size pipe on the oil return line.

After the carpet and backing are joined, the product is rolled at the end of the laminating range to await further processing. These in-process rolls were temporarily stored at various locations throughout the manufacturing area. The additional processing included cutting the rolled product into carpet squares and usually adding an adhesive and protective cover over it. The squares would be packaged into cardboard boxes, marked, stacked into a pallet load and moved to storage.

Nearby Range 6 was a similar piece of equipment, identified as "Laminating Range Number 400." This unit was also served by the main hot oil circulating loop and also attached backing material to rolls of carpeting. The third load loop on the hot oil system was the adhesive mixing vats located on the second floor of the adjoining building. The operating temperature and characteristics of this load were not available. The 500 gallon hot oil system expansion tank was located above the roof on this building.

BUILDING CONSTRUCTION

The complex was generally a single story, noncombustible building with some sections having a mezzanine or two-story offices (See appendix A). The single story sections were almost 30-feet tall and the complex had approximate dimensions of 585 feet (north-south direction) by 885 feet (east-west direction). The total horizontally projected area inside building walls was approximately 536,000 square feet and the total floor area (including mezzanine and 2nd floor) was approximately 580,000 square feet. Except for some small offices, the laboratory and the truck staging/shipping area, the fire caused complete structural collapse of the Complex.

In the Carpet Service Center (original 1968 building and 1970 addition), the roof construction was described as tectum board roof deck supported by unprotected steel bar joists. In the Live Oak/ Milstar Complex (1972 addition) the roof construction was metal deck supported by unprotected steel bar joists. Throughout all of the buildings, the roof's steel bar joists were supported by unprotected steel beams and columns. Firefighters reported that the roof covering was a gray material without a gravel covering; this was most likely a single ply membrane roofing material. Bar joist depth was in the range of 2 to 3-feet under the roughly 30-feet tall roof deck. The mezzanines and second floors were constructed of concrete on metal forms supported by unprotected steel beams and columns.

Exterior walls were predominately non-load bearing brick faced hollow concrete block. Some metal panel walls over the steel supports were used on the east side of the complex and in auxiliary buildings. The former exterior walls, which became fire separation walls between building additions, were often non-load bearing brick faced hollow concrete block with some walls constructed of only hollow concrete block. It is estimated that both 10 and 12-inch thick hollow concrete block was used in wall construction. There was no evidence that any of the block was filled or reinforced to improve structural stability under fire exposure.

Large openings were made in the former exterior walls as the complex expanded. These openings were 10 to 20 feet wide and 10 to 15 feet tall. Opening protection consisted of rolling steel, Class A, 3-hour fire doors. Unverified information suggests that along one of the separation walls most openings had these fire doors on each side of the wall. There were also several large, about 3-feet high by 8 to 10-feet wide, convey or openings through separation walls. Hooded, automatic deluge sprinklers reportedly protected these conveyor openings although the exact arrangement of this protection could not be documented. Interior sector firefighters said that none of the seven or eight fire doors they were able to reach could be completely closed. They indicated that many would partially close before stopping in the pinched guide tracks. At least one fire door could not be coaxed out of its holder located above the opening.

Building Fire Protection

The complex was protected throughout by automatic sprinklers, including in-rack sprinklers in a number of locations. (See Appendix C for automatic sprinkler design information). In addition, small hose stations with 75-feet of fire hose and adjustable spray nozzles were reportedly distributed throughout the manufacturing and storage areas. Twenty-six automatic sprinkler risers were supplied from a 12-inch looped fire protection water supply main around the Complex. The looped main also supplied yard hydrants spaced at about 300 feet intervals. (See Appendix A for site fire protection diagram). Each automatic sprinkler riser was controlled by a post indicator valve (PIV) near its connection to the looped main. In addition, the looped fire protection main was subdivided by PIVs to allow segments to be isolated along with the hydrants and risers in that segment.

Water supplies to the looped fire protection consisted of two separate city water main connections and two on-site fire protection water reservoirs, each with a diesel engine driven fire pump. The city water main connections were from public mains on the north and south sides of the complex and these public mains did not involve the same parts of the city water main grid. On-site water storage reservoir and pump combination "A" was a 500,000 gallon grade level tank supplying a 2,500 gpm, 125 psi diesel engine driven fire pump. Reservoir and pump combination "B" was a 400,000 gallon grade level tank supplying a 2,000 gpm, 125 psi diesel engine driven fire pump.

The city water main connection on the north side of the property was a 10-inch connection to the 10-inch circulating city water main on Swift Street. This city connection also supplied the automatic water fill system on each fire protection water reservoir and a 10-inch supply line directly to the plant's looped main. The city water connection on the south side of the property was a 12-inch connection to the 10-inch circulating city water main on Lukken Industrial Drive West. This city connection also supplied an 8-inch domestic service into the complex. (Water flow test data and city water system data from the day of the fire are attached in Appendix D.)

With the on-site fire pumps operating and the city water supply connections, over 8,500 gpm at 20 psi would have been available for fire protection needs. Over 6,000 gpm at more than 95 psi would have been available on the looped fire protection main for use by the automatic sprinkler systems.

Each automatic sprinkler riser appeared to have an outside water motor gong. Paddle type electric water flow switches on each sprinkler riser were supervised and monitored by a Pyrotronics System 3 fire alarm panel located in the guard house on the south side of the site. This panel was monitored by a central station fire alarm service and also directly connected to the Troup County 9-1-1 Emergency Dispatch Center.

Supervision of the outside PIVs was done with plastic breakaway tags which were reportedly checked during weekly recorded inspections. Although it could not be confirmed, LaGrange firefighters reported that some PIVs may have been closed at the time of the fire. The location of these PIVs could not be identified at the time of this investigation. Supervision and monitoring of the fire pumps, valving, and controls could not be determined because access to the pump houses was denied.

Regular testing and maintenance of the fire detection and alarm system was provided by the central station fire alarm service. LaGrange Fire Department indicated that the Milliken facility had frequent unwanted/false alarms from the fire alarm system.

Milliken officials indicated that there were several locations within the Complex where the evacuation alarm signal could be activated. This system was reportedly used to alert employees to the rapidly spreading fire conditions during the incident. It is believed that this system performed satisfactorily.

THE FIRE

On the morning of January 31, 1995 at about 10:11 a.m. the 9-1-1 Center was informed that the fire alarm service company would be testing the sprinkler system waterflow alarms. The 9-1-1 Center placed the system into a "do not respond" mode. A second telephone call was received at about 1:16 p.m. indicating that the testing was completed and to return the fire alarm to "service." About two minutes later, the 9-1-1 Center received a request from the Milliken plant's guard house to again place the fire alarm into a "do not respond" mode. Reportedly repairs were being started on one of the automatic sprinkler risers on the south side of the plant. This repair was described as a replacement of a section of pipe for the water motor gong.

Shortly before 2 p.m. on January 31, 1995, a hot oil leak began at a 1-1/2-inch diameter rotating coupling located on the second level of Laminating Range No. 6. The rotating coupling connects the fixed hot oil supply and return piping to the heated revolving rollers which compress the carpet and its backing together. Milliken reported a similar leak at another rotating coupling about a week before the fire and this leak was controlled without an ignition of the hot oil. The first leaking coupling was said to be located on the lower level of the range.

According to Milliken's Dean Jackson, who responded to the hot oil leak report from his office about 225 feet away, a massive white vapor cloud was spreading from the leak's location both upward and down to the floor. The base of this cloud was described as about 2 to 3 feet in diameter and the top about 10 to 15 feet in diameter. Emergency procedures direct operators to stop the hot oil circulation pumps within the subloop when a leak occurs. It is believed that this was done. Before Jackson reached the area, the vapor cloud ignited without a distinct noise from an undetermined ignition source. Flames were initially observed from the second level coupling leak down to the floor but they rapidly covered the area from the floor to the bottom of the roof deck.

Employees pulled racked fire hoses equipped with spray nozzles in an attempt to control the fire before the automatic sprinklers operated. Two nearby hoses were being used as automatic sprinklers began to operate. Dean Jackson indicated that a propane fueled lift truck quickly became engulfed in the fire and employees evacuated the area shortly thereafter.

The status of the main hot oil loop circulating pumps at the time of ignition and after was unknown. It is not known at which point during the fire these pumps shut down. No oil was recovered from the system after the fire.

The Troup County Emergency Communication Center (9-1-1 Center) records indicate that the fire department was dispatched at 2:02:28 p.m. in response to a telephoned fire report received from the Milliken plant's shipping and supply area. The plant's guard house telephoned at 2:04:54, also reporting a fire at the plant. Taped records from the 9-1-1 Center suggest that there was a plant fire alarm activation at about 1:51 p.m. but no response was sent because there was a hold on the directly connected fire alarm system. This suggests an 11-1/2-minute delay.

The Deputy Chief of Suppression was the Incident Commander (IC). He responded to the Complex from the Swift Street side and his initial size-up came from the intersection of Swift and Lindsey. He reported seeing heavy black smoke and flames pouring from the northeast area of the building's roof. The Swift Street plant entrance gate was locked but it was opened by Milliken employees as the IC was turning around to enter the complex. The first location of the Command Post was on the east side of the structure near the mid-point of the east wall (See Appendix E). The LaGrange Fire Department assigns sectors based upon the street address side being Sector A and working clockwise around the structure. Therefore, the Command Post was initially in Sector D.

A second alarm was requested at 2:12 p.m. which brought the third LaGrange engine to the scene and Troup County Fire Department units into LaGrange for standby. The County's response was believed to consist of two engine companies and a rescue company. A third alarm was requested at 2:14 p.m. which brought a West Point, Georgia (14 miles away) engine, two Troup County engines, a Troup County tanker and a portable breathing air cascade to the scene. At 2:18 p.m., the two off-duty LaGrange Fire Department battalions were recalled. At 2:28 p.m. another request was made for a ladder-platform from Carrollton, Georgia (50 miles away) and a ladder truck from Coweta County, Georgia (27 miles away). Records indicate that no alarms beyond the third were formally requested. Other firefighters were rotated into the scene from the involved fire departments to relieve the on-scene crews.

Arriving at the east side of the complex, the IC observed through the open loading dock doors flames extending from the floor to as high as he could see inside the building. He also reported that bright sparks, likely electrical, were falling from the ceiling area. About 10 minutes after arrival, the IC indicated that the fire had traveled to the south where flames were observed through windows at the top of the east wall over the maintenance shop area. These windows would be slightly south of the middle of the east wall.

A police "dash camera" video tape indicates that the fire had already broken through the roof in the northeast corner of the complex at the time of their arrival. The column of dark black smoke was already about 150 feet in diameter and several hundred feet in the air. Fire units also reported being able to observe the smoke column as they responded from Station 1. Weather at the time of the fire was sunny, temperature about 60 degrees Fahrenheit, and the wind was out of the northwest at about 10 to 15 mph.

Engine 4 (1,500 gpm with 500 gallon water tank) was operating in place of Engine 1. They entered the plant from Lukken Industrial Drive West (front side) and were directed by plant employees to the truck loading dock at the building's northeast corner (see Appendix E). A preconnected 1-3/4 inch line with spray nozzle was stretched to a personnel entrance door at this point. The crew reported flames extending from the floor up as high as they could see and 15 to 20 feet away from their position. They entered no more than 20 feet into the building and stayed long enough to empty the 500 gallon tank. As they were working, the engineer hand stretched a 3-inch supply line to a nearby yard hydrant and subsequently pulled a parallel 3-inch supply line. The crew abandoned the handline and placed into operation a deluge gun equipped with a 500 gpm adjustable spray nozzle on the ground near the truck dock. The deluge was initially directed onto the fire over the outside wall and then later used for exposure protection of adjacent equipment and storage vessels on the east side of the plant. These items included cooling system compressors, a large electrical transformer, small cooling tower, and two vertical storage tanks containing "PrudiWater."

Eventually this downwind position became untenable and Engine 4 was moved to Swift Street. The ground deluge, which was being directed by the crew, was set to a fixed position on exposed equipment and was disconnected from Engine 4 and attached directly to the 5-inch supply hose. After the move, Engine 4 pumped to Engine 3 while the crew concentrated on protecting the three dry bulk material storage silos at the northeast corner.

Because their primary vehicle was being serviced, the two member Rescue Company was operating reserve Engine 5, a 1,000 gpm unit. From Station 1, they responded from the east on the Swift Street side. On arrival, the crew observed that the fire had penetrated the metal deck roof at the northeast corner and the fire was running to the west against the wind along the northern edge of the roof.

Initially Engine 5 was positioned on Swift Street between O'Neal and Lindsey (see Appendix A for streets and Appendix E for placement) near a hydrant on the south side of Swift. Before completing their setup to use the deluge gun on the top of the engine, the fire had moved to the west past this position. The crew moved the engine to the north side of Swift just to the west of Lindsey. Using parallel 3-inch supply lines from the hydrant, they operated the 500 gpm spray nozzle equipped deluge from the top of the engine at the second position. They were unable to reach the fire with an effective stream from this position.

This attack was suspended when Engine 3 completed its reverse lay from Engine 4 to the hydrant that was supplying Engine 5. As the Rescue Company's officer helped connect Engine 3 to the hydrant, Engine 5 was moved further to the west and parked on Swift between Lindsey and Polk Streets. After assisting Engine 3 and while waiting further assignment, "explosions" began to occur along the plant's northern wall. Brick and hollow concrete block from the top 5 to 8 feet of the north wall would "explode" in 60 to 70 feet long sections with debris reaching the fence line about 60 to 75 feet away from the wall.

Subsequently (See Appendix E), Engine 5 connected to a hydrant near Swift and Polk Streets to supply a 3-inch line to Engine 3 and to use their deluge set which had been changed from a spray nozzle to a 1-inch straight tip. There was insufficient water available to supply both operations so they shut down the deluge set and continued to relay pump to Engine 3 through the dual 3-inch supply lines (a second supply line had been laid to Engine 3). This operation continued into the evening.

Engine 3, a 1,500 gpm unit, arrived about 20 minutes into the fire after completing another assignment which was received at nearly the same time as the Milliken fire. This engine was assigned to Sector C and entered the northeast corner of the plant from Swift Street. They assisted Engine 4 with setting-up the deluge and reverse laid a 5-inch supply line to a hydrant on Swift between O'Neal and Lindsey from Engine 4 (See Appendix E). They pumped the supply line and also operated an engine mounted deluge onto a roof mounted 500 gallon capacity hot oil expansion tank. The operation continued until Ladder 11 came to Sector C between 4:30 and 5:00 p.m. At that time, the engine was moved to allow the ladder to set-up and then engine supplied the ladder stream only.

Engine 2 entered the complex from the main entrance off to Lukken Industrial Drive West and initially staged at the southeast corner. They observed a Milliken employee stranded on the roof. While Ladder 11 attended to the rescue, Engine 2 reversed laid from the Siamese connection on the underground loop main to a city hydrant on Lukken and pumped into the Siamese connection. The engine crew became part of the interior sector operations which worked through personnel entrance doors along the south side of the building. Later in the fire, the engine also supplied water for other fire suppression operations in Sector A.

Ladder 11, a 100-feet rear mounted straight ladder, entered the complex from the main entrance off Lukken and initially staged behind Engine 2 near the southeast corner. Their first assignment was to rescue the employee stranded on the roof. The truck was pulled facing into the building at a location near the complex's southwest corner (See Appendix E). The rescue was completed using the main ladder. The firefighter who went to the roof to assist the employee down the ladder reported that heavy black smoke and billowing flames were observed pouring from the roof at the northeast corner.

After the employee rescue, the ladder picked-up and stayed for a brief period at the complex's southwest corner before being assigned to a fly pipe operation in Sector A. The truck was positioned facing into the building about 250 feet east of the southwest corner with the rear of the unit nearly at the south curb line of the complex's roadway. The ladder was extended about 60 feet at an angle of about 70 degrees. The operation initially started with an adjustable spray nozzle with a water supply from Engine 1 which had arrived later in the fire. This stream was barely reaching the building and the spray nozzle was changed to a 1-1/4-inch solid stream nozzle. While this improved the reach, the stream had little effect on the fire. This operation was subsequently discontinued and the ladder given another assignment.

The ladder was ordered to Sector B to establish another fly pipe operation at the north wall line of the 1968 original building. Several hundred feet of 5-inch supply line was hand laid to extend the water supply from Sector A to the new position. This time the ladder was backed into the building to a point just outside the collapse zone and the unit operated the 1-1/4-inch solid stream nozzle along this fire separation wall. Firefighters reported that the roof to the north of this wall had already partially collapsed before this operation started. The ladder operated for 30 to 40 minutes before being ordered to move due to an impending wall collapse. The ladder was still in preparation for being moved when a partial west wall collapse occurred, missing the truck.

At about 4:30 p.m., Ladder 11 was ordered to Sector C on Swift Street near the northeast corner of the complex. There they operated a fly pipe in this area until about 2 to 3 a.m. the next day when the operation was stopped for environmental sampling. During the unit's work at this location, the crew reported difficulty in extinguishing flames in this corner that kept flaming up after being knocked down.

Engine 1 arrived later in the fire staffed with the off duty battalions. The unit was initially positioned within Sector A (See Appendix E) and supplied Ladder 11 and interior sector handlines. After interior sector operations stopped, the engine briefly operated an engine mounted deluge from the complex's roadway near the southwest corner. Subsequently, the engine was ordered to sector C to supply other units in the area.

The Interior Sector was under the command of a LaGrange Lieutenant from the recalled off-shift battalion assisted by the Chief of the West Point Fire Department. His initial assessment of the 1968 original building's interior began shortly after his 2:30 p.m. arrival at the scene. At this time, smoke was reported high up under the roof with clean air at floor level. A crew of about five to six firefighters was assembled from a resource pool that numbered about 100 LaGrange, Troup County and other mutual aid department members. Another Lieutenant was assigned to log firefighters in and out of the building. A formal "passport" system was not being used.

The first hoseline was described as a combination of 2-1/2- and 3-inch hose that was supplying a 2-1/2-inch spray nozzle. They were also constructing a 3-inch supply line to a portable deluge set which was equipped with a 500 gpm adjustable spray nozzle. These were advanced into the structure, extended, and moved between open fire doors in the north and east walls of the original 1968 building. At one time, the handline was combined with the single hoseline into the deluge set for two supply lines. Later, the handline was broken off the deluge set and attached again to the 2-1/2- inch spray nozzle. These two lines were moved between four to six large wall openings where fire doors had failed to close. At each position, the line – either hand or deluge set – would operate for a time to knock-down the fire entering the door opening before moving to the next. Eventually, the Interior Sector Commander indicated that the flames coming through these open doors became more intense and movement between doors became more difficult as smoke banked down from the roof.

Movement of lines into the building and between the door openings was hampered by the storage racks and by the full pallet loads of finished product stored in the aisles between racks and at the end of rack structures. The Interior Sector Commander reported having to crawl over these contents to reach some of the nozzle operating positions.

Interior Sector operations continued for about 1-1/2 hours before crews were ordered from the building. Crews had been out of the building for only a few minutes when the eastern sections of the 1968 building's roof began collapsing. The fire had established itself into the last section of the complex and, with essentially no water supply for the automatic sprinklers, the total involvement proceeded rapidly.

Fire walls, with or without openings, are prearranged positions where the fire department can take a defensive stand to stop a fire's horizontal spread. Incident commanders should not rely on a fire wall to stop a fire's spread without some help from the department. Preplans should always anticipate that fire companies with hoselines, radios, and SCBA will need to be positioned along the unexposed side of a fire wall inside the building. Companies should also take positions on the roof next to the fire wall. Typically when fire walls are being relied upon, one or more fire protection systems or features have already failed and a major fire is in progress.

Even in fire walls that do not contain door or conveyor openings, the Incident Commander should anticipate penetrations of some form. In addition, fire walls can crack from the fire exposure and temperature differential between the fire exposed side and the unexposed side. These penetrations and cracks can allow flame and hot gases to ignite combustibles near the wall. Sometimes, heat conducted through the wall can ignite combustible storage stacked tight to the wall. Fire companies in position along the fire wall can observe these ignitions and take action to keep such fires small. They may also move or prewet storage to keep any fires from becoming established on the unexposed side.

When firewalls have openings for moving goods, equipment, and people, a potential path for fire travel exists. These openings may be protected by fire doors, fire dampers, water curtains or deluge sprinkler systems. All of these protection methods require regular maintenance and testing so they can be relied upon to work in an emergency. NFPA Standard No. 80 on fire door installation provides guidelines and cautions regarding the maintenance of the various fire door types. Some general precautions on the maintenance and testing of water based extinguishing systems are available in NFPA Standard No. 25. This can be helpful in preparing specific maintenance and testing programs for deluge sprinklers and water curtain protection. These two types of fire wall opening protection are highly dependent on having the water supply available at the needed flow rate and pressure. Sometimes the fire conditions are such that the water supply fails just when the opening protection is called upon to perform. This failure mode occurred at a number of conveyor openings during this fire.

Incident Commanders should also remember that tested and listed fire doors can pass flames and hot gases. NFPA Standard No. 252 on fire door testing, permits intermittent six-inch long flames on the unexposed side of the fire doors after 30 minutes of the fire exposure. These flames are permitted to be continuous during the last 15 minutes of the test. Besides small flames, fire doors transmit more heat than do fire walls and could ignite nearby combustible storage. In this incident, the fire door tracks were reportedly damaged sufficiently that the rolling doors would not completely close. This provided an even larger avenue for flames and hot gas spread.

Both diesel engine driven fire pumps stopped running during the early stages of this fire. Reports indicate that the first pump failed in under 30 minutes and that both pumps had stopped within 45 minutes of the fire's ignition. Both failures are believed due to a loss of cooling water to the diesel engines. There was no indication that either suction tank had been completely emptied at the time that the fire pump failed. This investigator was not permitted to check either diesel engine or to enter the pumphouses. The exact failure mechanism had not been identified at the time of this investigation.

The loss of the diesel fire pumps reduced the fire protection water supply volume by over 60 percent and the pressure by at least 50 percent. Without the pumps, the two city connections to the underground loop main provided all the water for fire protection. The fire department supplemented the two connections by pumping into the Siamese connection at the front of the plant. This probably did not offset more than 20 percent of the water supply that the two fire pumps provided.

In a facility with a fire pump, the owner often designates an employee to respond to the pump during emergencies to make sure the pump is running. Usually, alarm and running signals on the fire pump's condition are also transmitted to the fire command center. For diesel driven fire pumps, these signals often include temperature and oil pressure information. As part of its preplan, the fire department should incorporate the facility's emergency procedures into its plan. With reduced fire-fighter staffing, facility resources could be used in the effort to monitor vital fire protection equipment. If fire department resources permit, the Incident Commander should designate a firefighter with a radio to check on the fire pump because of its importance to fire suppression functions. There was no indication that the two fire pumps were attended to during this incident.

The Milliken & Company plant was a large structure with a variety of fire hazards consisting of high piled and rack storage, high temperature equipment, combustible liquids, and hazardous chemicals. In addition, outdoor equipment, bulk raw material storage silos, and fuel storage on the east side of the building represented additional special hazards. The combination of hazards overwhelmed the facility's primary fire defense in the area of origin, the automatic sprinkler systems. Problems with the fire wall opening protection in combination with the fire department's inability to mount sufficient support at the walls, resulted in the extensive loss.

Some exterior fire attack continued at the north and south sides of the complex by the use of two ladder streams and some deluge sets. At about 10 to 10:15 p.m., the two connections to the city water mains feeding the plant's looped fire main were shut-down. This action essentially stopped the water flow from the 26 broken and collapsed sprinkler risers. Before these actions were taken, a city utility official reported that the 15-inch diameter sanitary sewer outflow from the east side of the complex was completely full.

Incident Command was passed off from the Deputy Chief of Suppression to one of LaGrange's Captains at about 7:00 p.m. After a short meeting, the Deputy Chief assumed the duties of liaison between the fire department and Milliken officials involved with air monitoring of the smoke. The Incident Command and Milliken liaison changed again at about 7:00 a.m. the next morning. The fire was not declared under control until after February 1, 1995. Heavy equipment was brought to the scene to knock down walls and excavate smoldering carpet rolls and piles of carpet squares. Fire department units were operating at the scene around the clock for over a week, extinguishing spot fire as materials were removed by outside contractors.

LAGRANGE FIRE DEPARTMENT

The LaGrange Fire Department is divided into Suppression and Administrative Divisions, each commanded by a deputy chief. The Suppression Division operates three engine companies, a truck company and a rescue company out of three stations using a three platoon (battalion) system of 24 hours on and 48 hours off. Engines 1 and 2 and Truck 11 are each normally staffed with an officer and three firefighters. Engine 3 normally has an officer and two firefighters. The rescue company's normal staffing is an officer and one firefighter. Typical department strength is 17 on-duty personnel. There are three reserve engines maintained by the department.

The Administrative Division consists of the Fire Prevention Bureau, Public Education staff, Training function, and the Supply and Maintenance group. The Fire Prevention Bureau is also responsible for fire investigations. One member of the Public Education staff also fills the role of Public Information Officer during incidents such as this. Maintenance and Supply includes within its responsibility the maintenance and testing for the fire hydrants in LaGrange.

The normal response to a report of a structure fire is two engines, the truck company and the rescue company. The Deputy Chief of the Suppression Division will also respond during usual business hours and to confirmed fires outside this time. Engine 2's quarters are the closest to the Live Oak Complex and would usually be the first unit on the scene. At the time of the fire, all on-duty units were at Station 1 for drill and training.

Emergency medical service is provided by a separate service through contracts with the county. The ambulances are located at the hospital rather than in city fire stations. An EMS unit reportedly responds with the fire department to structure fires.

Radio frequencies used by the LaGrange Fire Department are a city-wide dispatching and operations frequency and a lower powered fireground frequency. The Troup County Fire Department uses a different dispatching and operations frequency. Most city and county fire units reportedly could use any of the three radio frequencies. There were no reports of radio traffic congestion resulting in difficult or lost communications.

BUILDING CODES

At the time of the fire, the 28,000 population City of LaGrange had adopted the 1991 editions of the Standard Building Code and The Standard Fire Prevention Code both published by the Southern Building Code Congress. The community also adopted the 1988 edition of NFPA Standard 101, The Life Safety Code published by the National Fire Protection Association.

LaGrange Fire Department reported that the Fire Prevention Bureau began performing new construction plan reviews in about 1984. Previously, new construction plan reviews were entirely done by the LaGrange building official. Representatives of Fire Prevention Bureau indicated that they have been attempting to increase their involvement in the plan's review process since that time. They also report attempting to perform more frequent fire prevention inspections of properties throughout the community. However, it was also indicated that the last time Milliken's Live Oak/Milstar Complex had been thoroughly visited was over 5 years before the fire and the last company in-service visit may have been several years before the fire. The manufacturing equipment initially involved in the fire may not have been observed by anyone in the fire department prior to the fire.

LESSONS LEARNED

1. **Automatic fire sprinkler systems will not control all fires involving preheated combustible liquids nor will they prevent the rapid collapse of lightweight unprotected construction exposed to rapidly developing fires.**

   Plain water automatic sprinkler systems are not always effective on fires that involve spills of combustible liquids, especially those that are heated above their flashpoint. The quality of liquid spilled and the type of surface the liquid falls onto often influence the success or failure of the fire to be extinguished. Incidents that continue to discharge combustible liquid, especially heated ones, reduce the likelihood of success of full fire extinguishment. Heated liquids spilled onto cooler floor surfaces, such as concrete on ground, cools the liquid, reducing the hazard. Spills onto porous surfaces, such as cardboard and carpet material, allow the liquid to "wick", improving vaporization and increasing the hazard. Combustible liquids heated above 212 degrees Fahrenheit present additional challenges because of their ability to quickly turn water into steam and to rapidly destroy fire fighting foams covering their surface. Combustible liquids also have the ability to float on the water's surface and this can cause the fire to spread to adjacent combustibles.

   In this incident, the fire reportedly also involved a propane fueled lift truck while employees were using small hose streams. This likely resulted in the release of propane gas which intensified the fire even though no rupture of the gas cylinder was reported. It is doubtful that the automatic sprinkler discharge was able to absorb much of the heat being released by the burning oil and propane gas.

   The building's lightweight, unprotected steel roof construction was unable to resist the fire and began to fail. The collapse of the roof structure most likely caused the failure of large supply mains for two automatic sprinkler systems which were located in the area of origin.

2. **Process controls must include procedures to effectively manage expected equipment failures. In addition, employees must receive training and practice in the implementation of these procedures.**

   Limiting the amount of combustible heat transfer fluid that can be spilled is critical to controlling this type of incident. The operating conditions of the thermal transfer system are such that the fluid was operating above its flashpoint and would be easily ignited. Milliken reported that the hot oil subloop circulation had been stopped but it was unknown if the main loop's circulation was still running. In addition, stopping the circulation alone may be insufficient to control the oil release and other means to reduce the amount of leaking oil, such as emergency drain down, may be needed. Fire officials need to review the process controls with plant officials to identify potential liquid (or gas) fuel sources and understand how to stop accidental releases. Preplans must incorporate the location and identification of these controls and emergency features such as drains.

3. **When automatic fire alarms are placed out-of-service, the plant should institute pre-arranged procedures to assure that real emergencies will be promptly reported to the fire department.**

In this incident, considerable delay occurred in the response of the LaGrange Fire Department because of the fire alarm system's status at the Troup County 9-1-1 Center. Pre-arranged procedures for reporting real alarms should have been established and the reason for taking the fire alarm out-of-service understood and monitored by a responsible person at the plant. For example, if the alarm system was removed from service to work on automatic sprinkler riser 6, then an alarm from riser 10 should result in an immediate emergency response because this alarm would be unexpected and likely be an indication of a real emergency. Indications are that such procedures either did not exist or were not followed in this incident.

4. **Maintenance and repair of fire doors must be performed on a regular basis and damage that affects the door's operation should be promptly corrected.**

Most of the fire doors protecting large openings in the several fire separation walls within this complex failed to close either automatically or manually. The overhead rolling steel type fire door utilizes a "U" shaped channel on each end of the interlocked slots that holds the door to the wall and guide it to the floor. These channels can be pinched closed such that the door will not close or they can become misaligned enough that the door will hang-up as it falls toward the floor. Milliken reported performing weekly checks of fire protection systems and the fire door operational deficiencies were brought to their attention several weeks before the fire. Apparently repairs had not been completed.

5. **Pre-incident planning must include the preparation of fire equipment and staffing response needs for each of several alarm levels. In addition, the mutual aid fire departments involved should conduct regularly scheduled drills to become familiar with multiple department operations.**

In operations of this magnitude, it is imperative that the incident commander have prearranged equipment and staffing resources available and understand the response characteristics of these resources. It is difficult to develop and implement an effective tactical plan without the knowledge of available resources. The capabilities of these resources should be identified through drills so they may be incorporated into the incident commander's tactical plan. Good preplans should include alternate tactics that might be needed if a built in fire protection feature is not in-service or has failed. In this fire, the automatic sprinkler systems failed early, yet support of the fire separation walls did not appear well planned.

6. **Regularly scheduled fire prevention inspections and review of plans for construction and major remodeling should be conducted and the information incorporated into the pre-incident plan.**

One purpose of fire prevention inspections is to check the operational condition of active fire protection systems and passive features such as fire doors. In addition, changes to plant equipment, operations, and hazards can be identified and incorporated into pre-incident planning activities. In this case, it had been approximately a year since an inspection had been made. The Milliken facility was both large and complex in terms of physical characteristics and it contained significant production and process hazards. In addition, the carpet storage presented serious fire suppression challenges to both the automatic fire sprinkler systems and fire department operations. A fire nearly anywhere in the complex would have represented a significant challenge to even a well prepared fire department.

7. **Through fire company in-service visits and pre-incident planning, the fire department leadership and plant officials should become familiar with each other's resources, operating procedures and capabilities. This includes the plant's active and passive fire protection features (automatic sprinklers and fire separations for example).**

Pre-incident planning, especially in a complex manufacturing facility, involves not only the typical physical objects such as sprinkler systems boundaries, control valve locations, hydrants, fire pumps, alarm systems, and fire doors, but also for senior officers and chief officers to become familiar with plant managers of key departments such as engineering, maintenance, and production. Respect and candor on both sides is important to successful outcomes during emergencies. The time for the fire department and the property owner to become acquainted with each other is not as the facility is burning to the ground.

For example, there was apparently confusion regarding the contents and hazards represented by the vertical and horizontal storage vessels located near the northeast corner and along the east side. Keeping contents and hazards information up-to-date may be difficult for the fire department but not for the plant. When both parties are familiar with each other, rapid and candid information exchange can occur that can be relied upon by all involved. Such information might have been proprietary and the plant reluctant to disseminate it before an emergency but having it during the incident from a reliable source can often be sufficient for fire department needs. Furthermore, it is difficult for fire department officers to be intimately familiar with each plant. Therefore, a pre-arranged plant contact is a tremendous resource during an emergency.

8. **Failed and collapsed automatic sprinkler systems need to be promptly isolated from water supplies in order to maintain water to viable automatic sprinkler systems and fireground operations.**

Fire protection water supplies for a large complex are not usually capable of supporting multiple sprinkler system operations. The large number of broken and damaged risers diverted water from viable systems and from fire department suppression operations. Outside post-indicator-valves (PIV) were available to shut-off individual systems or to isolate or sectionalize parts of the underground fire main loop. A plan showing each sprinkler system boundary and the control valve was posted in the plant guard house in Sector A. This information in combination with the PIVs could have been used to effectively control important water supplies. This is also another example of the importance of pre-incident planning and developing alternative tactical plans for fire protection system failures.

9. **The diesel engine driven fire pumps failed early in the fire most likely due to the loss of cooling water circulation through each unit's heat exchanger.**

Due to the number of collapsed and broken automatic sprinkler systems, the amount of water leaving each fire pump was likely beyond the usual operating range. This placed additional stress on the diesel engine and it is believed that an insufficient amount of cooling water entered the heat exchanger, resulting in the diesel engine over-heating. During an emergency, the plant should dispatch knowledgeable personnel to attend to fire pumps and the fire department should also consider sending firefighters to act as liaisons. The loss of the pumps reduced the complex's fire protection water supply by over 50 percent.

10. Large conveyor openings protected by deluge automatic sprinklers can lose protection when the water supply fails leaving an easy path for fire spread.

The protection of openings in fire separation walls by deluge or automatic sprinklers should be noted during pre-incident planning. Failure of piping or water supplies leaves these openings unprotected by usual means. Fire department tactics should include sending a crew with hose-line to cover these openings in order to prevent fire spread. In addition, it would be appropriate to check all fire separation wall openings to be certain doors have closed the combustible stock nearby have not ignited. All types of fire doors allow the transmission of considerable heat and smoke and even small amounts of flame are permitted around door edges. Nearby combustible stock or construction can be ignited by the heat and flames. Routine monitoring of all openings and penetrations in the separation walls must be part of the Incident Commander's tactical plan. Even when the doors, deluge sprinklers or fire stopping performs within Standards, these are weak points through which fire can spread. Blank fire separation walls should also be monitored for fire penetration and even standby hoselines advanced to key locations.

APPENDIX A

General Site Arrangement and Addition Dates
Fire Protection Water Supply

Appendix A (continued)

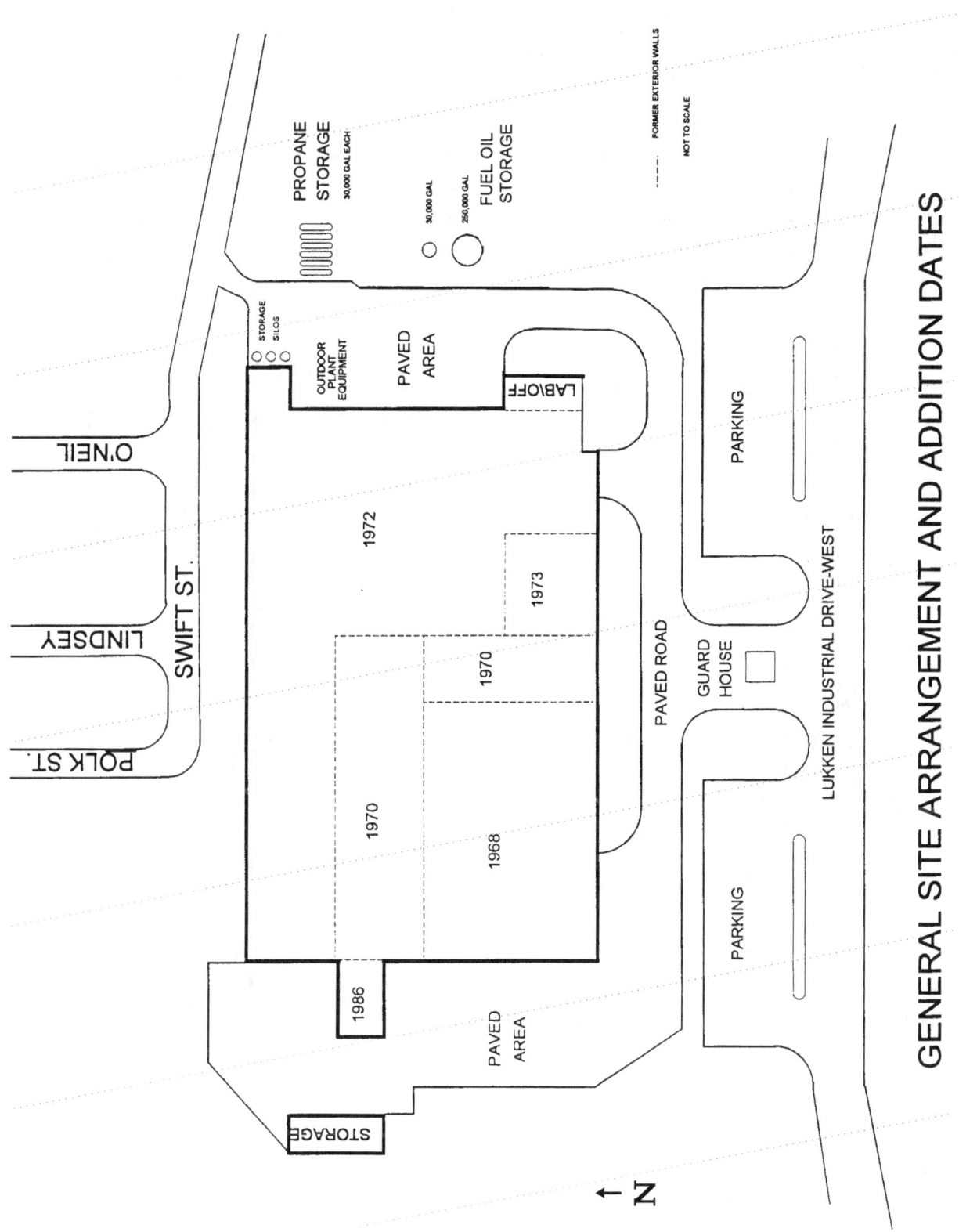

GENERAL SITE ARRANGEMENT AND ADDITION DATES

Appendix A (continued)

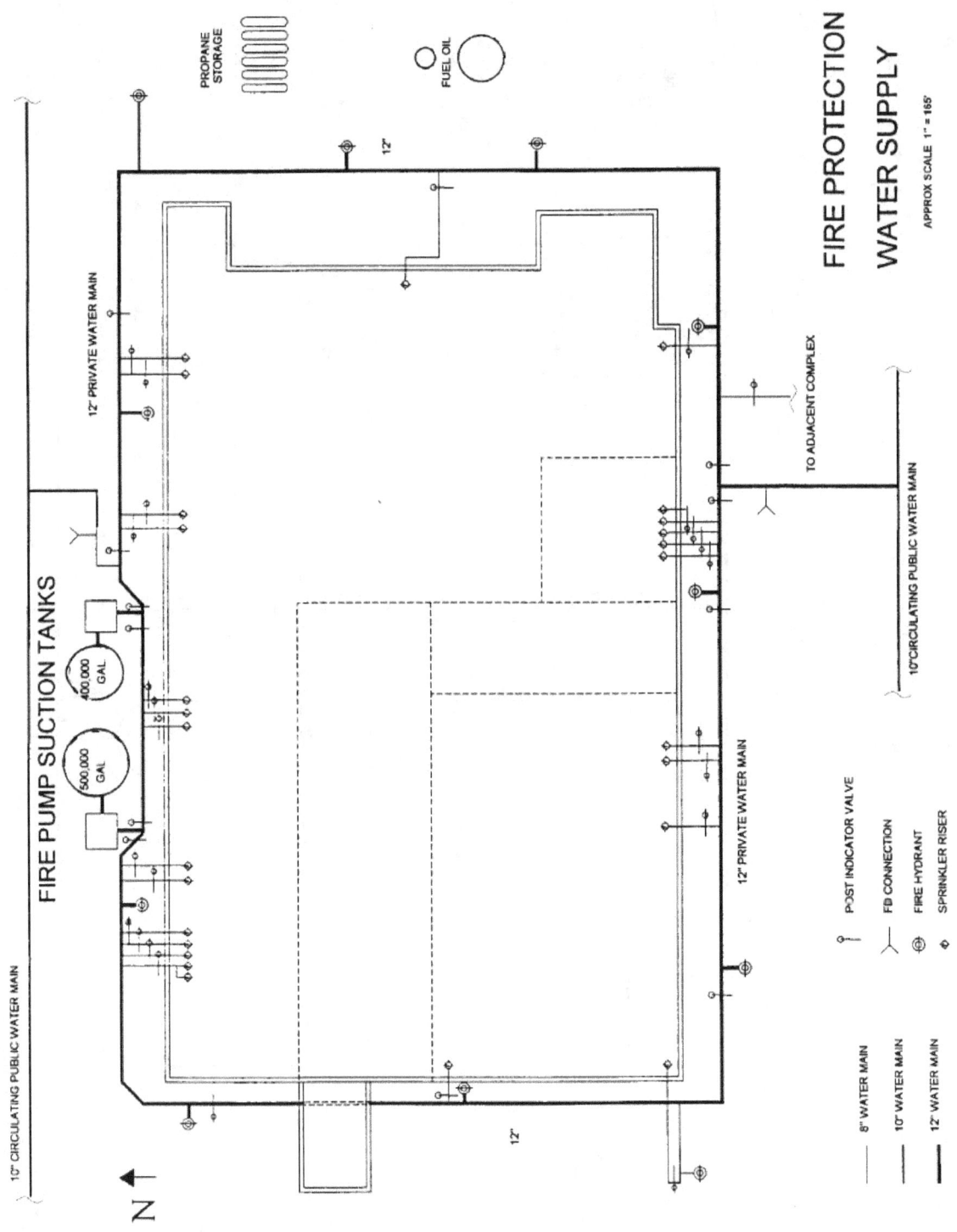

FIRE PROTECTION
WATER SUPPLY
APPROX SCALE 1" = 165'

APPENDIX B

Low Pressure Heat Transfer Systems

The most common means to provide heat for manufacturing and process needs is to use hot water and steam when direct fired heating is inappropriate. Water's heat capacity and heat of vaporization make it an excellent choice for heat transfer purposes. (These are the same features which make water an excellent extinguishing agent). Additional benefits of water are its low toxicity and low fire hazard compared to other heat transfer systems. Unfortunately, as manufacturing temperature requirements exceed about 250 degrees Fahrenheit, the heat transfer system pressures increase significantly as the temperature rises. In order to reduce the system pressure, a higher boiling point heat transfer fluid is substituted for the water. There are several manufacturers of over 40 different heat transfer fluids, some of which are combustible liquids.

In the typical closed loop liquid phase heat transfer system, components consist of the heater, circulating pump, expansion tank and heat consuming loads. An emergency drain or liquid storage tank is often provided for other than small heat transfer systems. The individual heat consuming loads may also have independent circulating pumps and different heat transfer fluid temperature needs. In this case, hot transfer fluid from the primary or supply loop would be blended with cooler fluid in the load loop. An equal amount of fluid from the load loop would be returned to the supply loop for reheating. The operating temperatures of liquid phase heat transfer systems can exceed 600 degrees Fahrenheit.

In addition to the elevated temperature, fire hazards include the combustible nature of many heat transfer fluids. These fluids are made from synthetic hydrocarbon mixtures, mineral oil, paraffin oils, silicone oils, and other hydrocarbon materials. Sprays, mists, and vapors released from the closed systems can create an immediate fire hazards which can be ignited by surrounding equipment. One ignited, intense, smoke fires occur with the heat transfer fluid as fuel. Stopping fluid flow is critical to successfully controlling and extinguishing the fire. Fluid fires can also ignite nearby combustible materials and can result in early collapse of the building's structural system.

A special fire hazards exists for the heaters used in closed loop liquid phase heat transfer systems when combustible fluids are used. These heaters are often direct fired using open flame burners. A liquid leak inside the heater is immediately ignited and these fires have resulted in substantial losses. Special fire detection and suppression systems can be employed by the most common loss prevention measure is the physical separation of the heater from the rest of the property.

Emergency procedures and operating controls must be prepared for handling heat transfer fluid leaks. Immediately stopping the flow of fluid and perhaps system drain down can be used to limit the amount of fuel available for a fire. Procedures also need to consider the hazards resulting from the interruption of the heat transfer system.

continued on next page

21

Appendix B (continued)

In this incident, Milliken indicated that the liquid phase heat transfer system used Therminol 55 ® synthetic heat transfer fluid. According to manufacturer literature, it is a synthetic hydrocarbon mixture with a clear yellow color. It has a flash point of 350 degrees Fahrenheit and a fire point of 410 degrees Fahrenheit. The reported autoignition temperature (ASTM D-2155) is 675 degrees Fahrenheit. Therminol 55's recommended operating temperature range is 0 degrees Fahrenheit to 550 degrees Fahrenheit. Overheating or fluid contamination may result in the precipitation of solids from the liquid. These solids can damage equipment, plug valves and pipe lines, and interfere with heat transfer from and to the fluid.

® Registered Trademark of Monsanto Company

APPENDIX C

Automatic Fire Sprinkler Design Information

The following automatic sprinkler design information was available from the LaGrange Fire Department:

Area	Density gpm/sq ft)	Area of Application (sq ft)
1968 Original Building	0.50	5,000
1970 North Addition	0.40	7,200
1970 East Addition	0.40	7,200
1973 Warehouse Addition	0.40	7,200

The 1972 manufacturing plant addition was originally protected by Ordinary Hazard pipe schedule systems following the requirements of NFPA Standard No. 13 on automatic sprinklers. Modifications or reinforcement of these sprinkler systems could not be determined during this investigation.

Water Supply Information
City of LaGrange Water Utility
LaGrange Water Department Delivery Pressure Chart

WATER SUPPLY INFORMATION

Waterflow test data for Lukken Industrial Drive West:

	Static	Residual	Flow	Flow@20 psi
August 30, 1994	108 psi	70 psi	1200 gpm	1900 gpm
May 24, 1993	96 psi	82 psi	2700 gpm	3300 gpm

The on-site fire protection water supply:

	Rated	150% of Rated
Fire Pump A:	2,500 gpm@125 psi	3,750 gpm@81 psi
Fire Pump B:	2,000 gpm@125 psi	3,000 gpm@81 psi
	4,500 gpm@125 psi	6,750 gpm@81 psi

CITY OF LAGRANGE WATER UTILITY

Water is obtained from the Chattahoochee River and treated before entering the city's water delivery system. At the treatment plant, a total pumping capacity of about 13,150 gpm is available using all high service delivery pumps. Additional flow is available from three elevated water tanks ranging in storage capacity from 1 to 2 million gallons each.

Records indicate that during the afternoon and evening of January 31, 1995 four of the five pumps were placed into service and that nearly 10 million gallons of water was delivered that day. This figure was significantly above the normal daily consumption for the time of year. The delivery pressure chart recording at the treatment plant documents the pressure drop at the start of the fire and a spike when the city water connection to the Milliken plant was closed. During the period between these events, the treatment plant delivery pressure remained within the normal operating pressure range of 72 to 96 psi. This indicates that the water mains in the area of the fire were delivering water at their maximum rate.

Appendix D (continued)

LAGRANGE WATER DEPARTMENT DELIVERY PRESSURE CHART

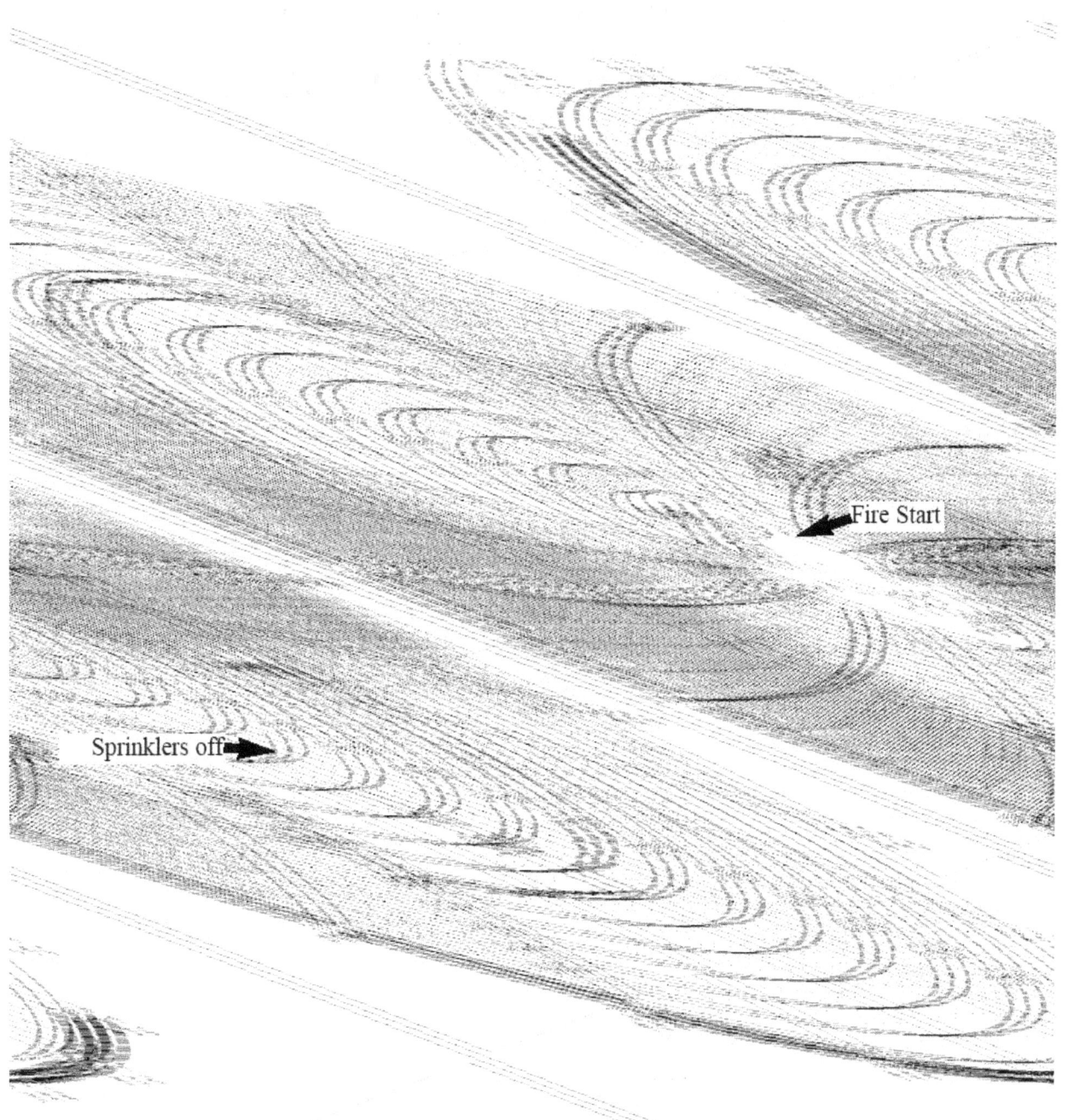

APPENDIX E

Early Fire Apparatus Positions

Later Fire Apparatus Operations

Near Ending Fire Apparatus

Appendix E (continued)

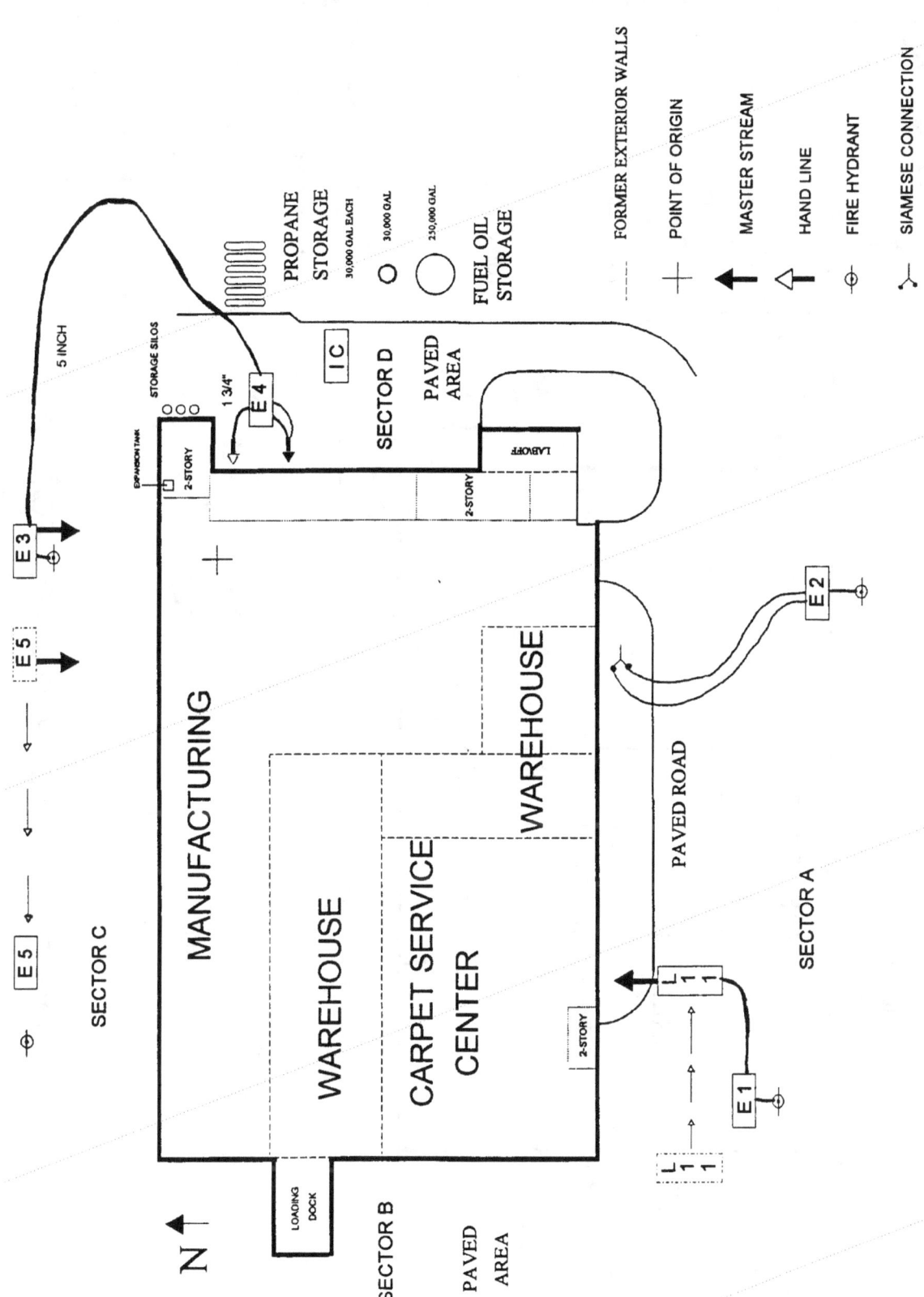

APPENDIX E   EARLY FIRE APPARATUS POSITIONS

Appendix E (continued)

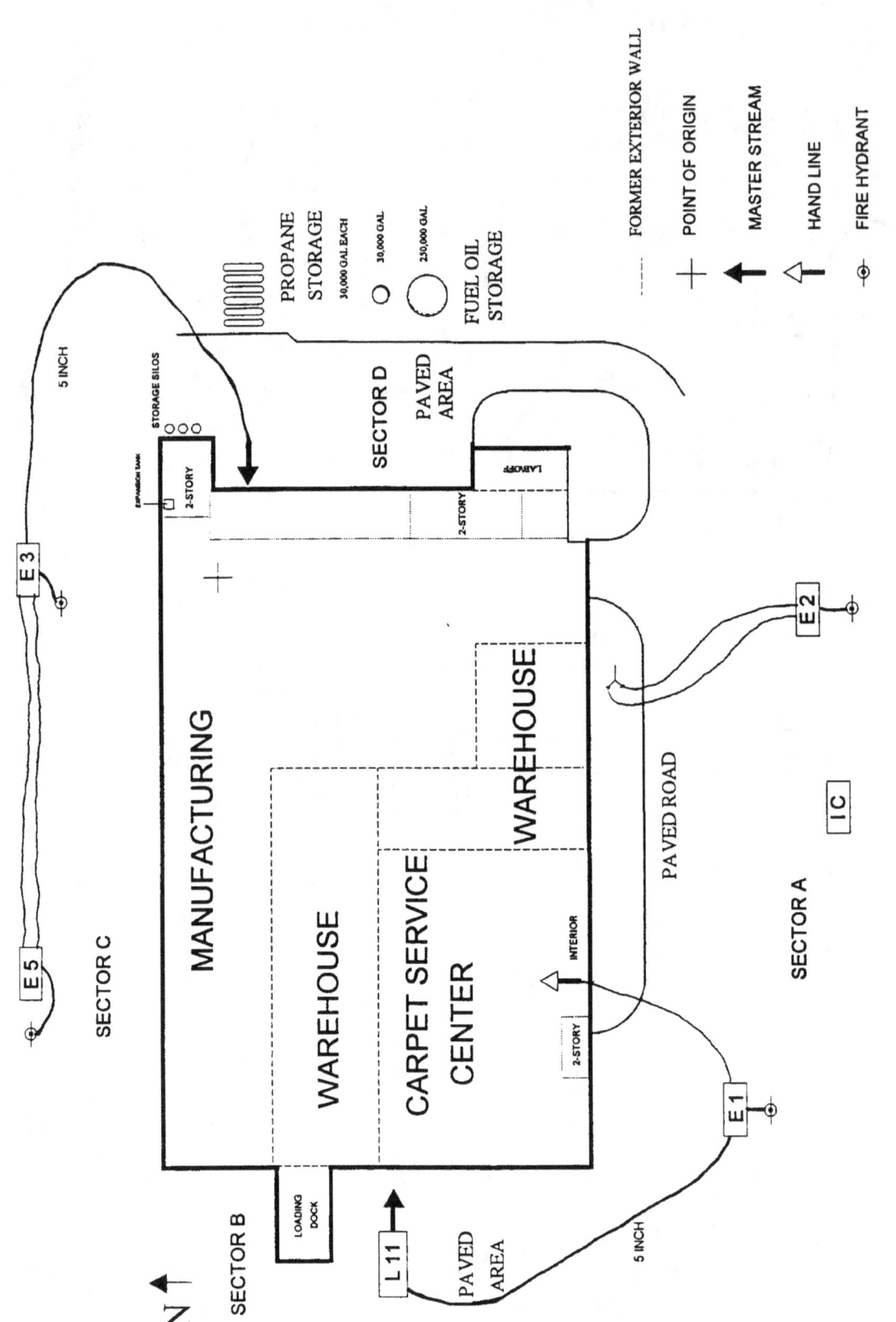

APPENDIX E LATER FIRE APPARATUS POSITOINS

Appendix E (continued)

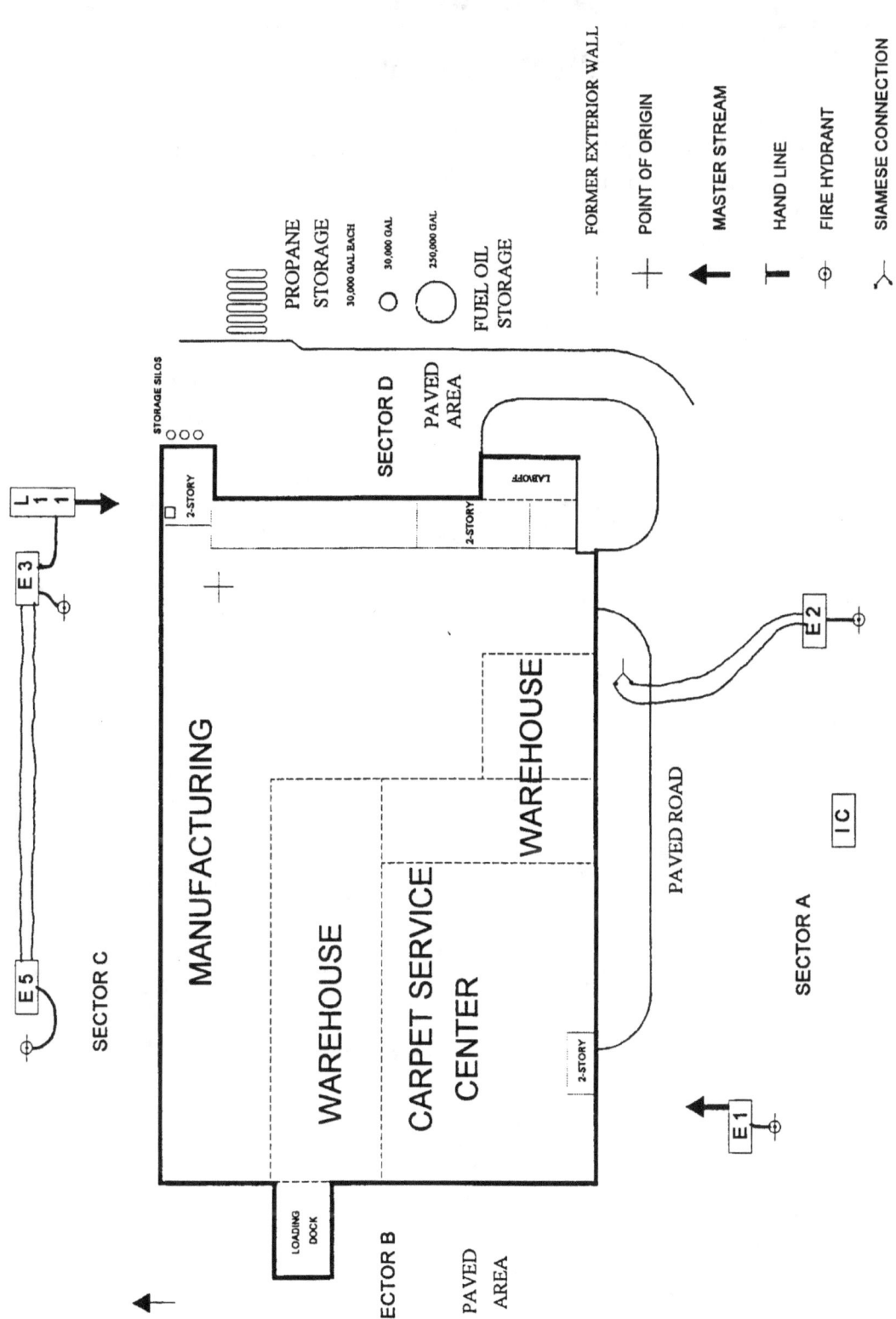

APPENDIX E   NEAR ENDING FIRE APPARATUS POSITIONS

APPENDIX F

Photographs

Photographs 1 to 10 were taken by the LaGrange Fire Department.
The balance of the photographs were taken by Thomas Miller.

Appendix F (continued)

Photo 1. Point of initial fire department attack in the northeast corner of the 1972 addition. The first line entered the building at a location near the photograph's center between the truck trailer and the building.

Photo 2. The location of Incident Command on the east side of the building. The truck trailer with the star (above the car's warning lights) is in photograph 1.

Appendix F (continued)

Photo 3. Conditions along the north side of the plant looking from west to the east. The photograph was taken very early in the fire and represents conditions not long after the fire department arrival.

Photo 4. One of the early master streams on the fire on the north side. There is an elevation drop of 15 to 20 feet between street level and ground level at the brick wall. One of the heat transfer fluid boilers is in the background just off the engine's tail board.

Appendix F (continued)

Photo 5. Conditions along Swift Street northeast of the plant. The large diameter supply line is in place to supply Engine 4 operating in the corner.

Photo 6. The two heat transfer fluid boilers and support equipment that supplied the carpet making range initially involved in the fire.
Engine 4 is behind the concrete embankment.

Appendix F (continued)

Photo 7. Conditions on the south side of the complex looking north from a position out from the northeast corner. At this point, the fire involves most of the 1972 addition.

Photo 8. Condition on the south side of the complex looking north from a position out from the northeast corner. At this point, the fire involves most of the 1972 addition.

Appendix F (continued)

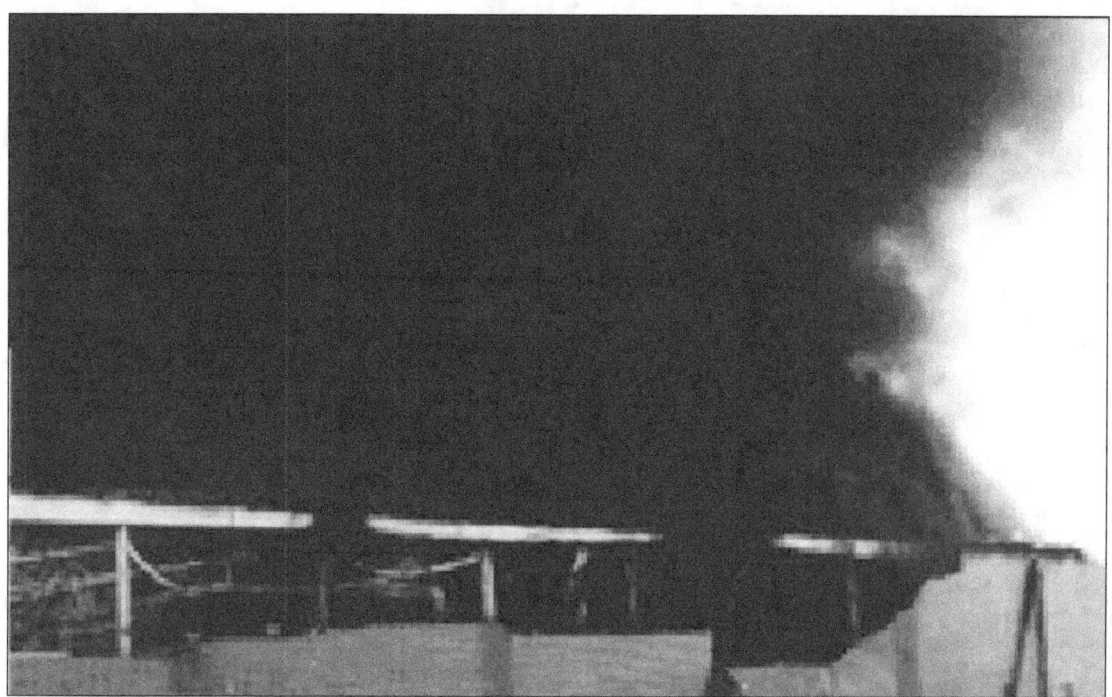

Photo 9. Conditions at the northwest corner of the building looking west from near Swift and Polk Streets. The outside post indicator valves (PIVs) and outside flow alarm bells for the sprinkler risers can be seen.

Photo 10. Close-up of the area in photograph 9. In addition to the flow alarm bells and PIVs, the top of a yard hydrant is in the lower left corner.

Appendix F (continued)

Photo 11. Several days after the fire, the northeast corner of the complex. The LP-gas storage tanks, raw material storage silos, and heat transfer fluid boilers and support equipment are shown. Taken from Swift Street looking west.

Photo 12. Nearly the same position as photograph 3. One of the fire protection water storage tanks is in the right hand side of the photograph. Taken looking east from near the intersection of Swift and Polk Streets.

Appendix F (continued)

Photo 13. Some of the same area in photograph 4 but from a different angle and after the fire. In the bottom right hand corner is one of the heat transfer fluid boilers.

Photo 14. The east fire pump house and a view into the building showing some of the manufacturing equipment and ductwork. The west side of the heat transfer fluid boiler in photographs 4 and 13 is on the left edge of the photograph.